SUGGESTION SCHEME

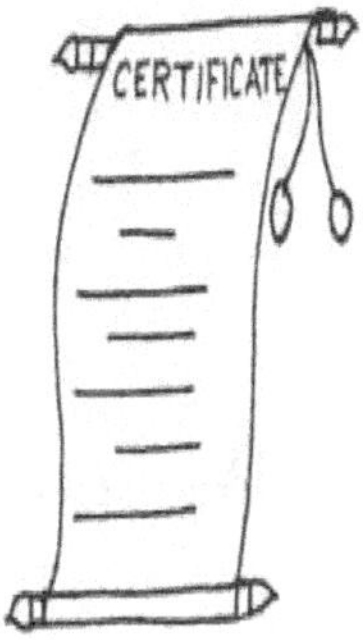

CERTIFICATE

CASH

A HANDBOOK

XXXXXX
TROPHY

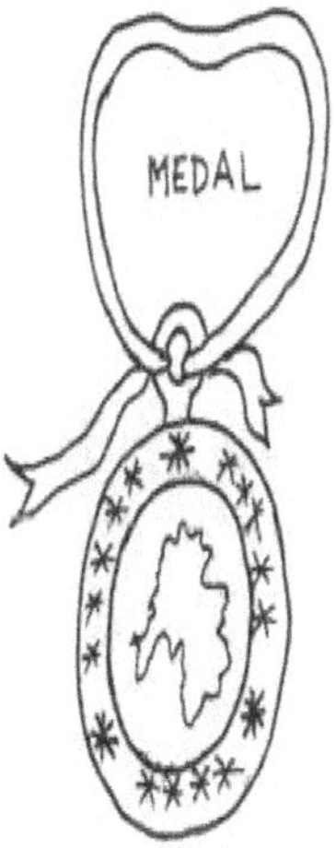

MEDAL

DINESH CHANDRA

SUGGESTION SCHEME

(A Corporate Handbook)

ISBN :: 9798720406561

Dinesh Chandra

PREFACE

In the present competitive scenario, continuous improvement is the need of the day. Survival is the fittest and so the growth is most sought for. This is true for all organizations – small or big.

"Suggestion Scheme" is an ultimate tool that support the need for overall growth and hence this book.

This handbook, in chronological order, provides guidance to prepare, start, and run the scheme throughout the lifespan of the organization.

The other features include few examples of suggestions, do's and do not's, sample forms, certificate, examples of suggestions etc.

In-fact, it is a handy reference manual to run the scheme and enjoy the benefits.

In a nutshell

"The suggestion scheme is a boon to all organizations irrespective of size, and or product. Its' benefits are beyond imagination and run parallel to the scheme throughout the lifespan of organization itself. "

Dinesh Chandra

I-N-D-E-X

✳✳✳✳✳✳✳✳✳✳✳✳✳✳✳✳✳✳✳✳✳

Introduction

The successful participant of suggestion scheme is provided with cash reward, medal and or certificate. This leads to recognition of participant in the section, department, division and the organization too. This recognition can also be linked to employee growth / promotion depending upon the contribution of the suggestion. Sometimes, in the process, it so happens that one participant becomes the ICON of the organization and thus a real morale boosting of employee in the true sense of the term, across the company.

But the hidden benefit of the suggestion scheme is that it keeps the organization healthy, from all angles, throughout the life span of suggestion scheme. The growth of the organization depends on successful running of the scheme.

On the contrary, the scheme being the most popular, collective and cheap in nature, it is very difficult to maintain and keep rolling for longer time for the various reasons as mentioned in do not's section of this handbook. Thus, do's section of this handbook plays a significant role for running the scheme. The full systematic support and approach of the top management for this scheme is essential to reap the benefits for organization for longer period of time. There is no end to the scheme and so the benefits.

✸✸✸✸✸✸✸✸✸✸✸✸✸✸✸✸✸✸✸✸✸✸✸

The Scheme

On all topics relevant to organization, from all level, a suggestion is invited, received, acknowledged, evaluated, tried-out, implemented, appreciated and awarded in a fair manner.

The Top Management, Human Resource and Finance departments play a key role in supporting and monitoring the scheme through out while departmental and divisional structure involve themselves in processing the suggestions in a faster and fair manner. Generally, all the roles in this context including coordinator, departmental head, divisional head, evaluating team, testing team and implementing team are clearly defined / laid down in the suggestion manual in detail as suitable to organization with FAQs.

Coordinators are selected at all levels like department, division and company level. They in-turn report to their respective leaders i.e. Departmental Head, Divisional Head and Company Head. Each Division will have evaluating team comprising of coordinators, selected members and leaders included. The final approval authority lies with departmental head, divisional head or the company leader as the case may be, in other words depending on the benefits. In case of expert opinion, the suggestion is marked to respective expert in the organization, irrespective of

origine/department of suggestion. Similarly, if the suggestion pertain to other division, the same is sent to respective division for their evaluation too. All the suggestions are given a top most priority for faster processing and secondly to hide the identity there is no name of the applicant on the form but only a serial number is maintained.

✳✳✳✳✳✳✳✳✳✳✳✳✳✳✳✳✳✳✳✳✳

The Conditions

The first and the foremost condition for smooth and successful running of a suggestion scheme in any organization is that it must have an illustrated and comprehensive manual to be followed by all and one in the organization. The presence of committed leaders serves as the core of the scheme. A few conditions, for easy reference, are being mentioned below.

1. Participation :: The participation is expected from each and every member of the organization irrespective of their post / level.

2. Fair approach :: The complete procedure followed must be fair on each and every step including registration, evaluation, testing, implementation and awarding.

3. Monitoring and Follow-up :: Each and every suggestion must be monitored and followed up by top management through coordinators and leaders till it is implemented. A digital approach / update is appreciated.

4. Evaluation :: Prompt evaluation of suggestion is the key to success of Scheme. Any delay will loose the benefit aimed at. Each and every recording must be dated.

5. Support :: Support to all junior participants must be provided by senior staff and leaders so that they can present their thoughts in a better way with clarity.

6. Motivation :: Participant must be motivated for taking part in the scheme. The first and formost requirement of motivation is appreciation for participation from senior.

7. Guidance :: All suggestor must be provided with proper guidance at all time.

8. Training :: Each and every employee must be trained on   necessity of suggestions, quality and efficiency improvement,   cost   reduction, Productivity, Safety. Etc.

****

Preparation

As the participation is expected from each and every member of the organization, the quantum of work becomes huge and need precision at each and every step of process. Now, it is very much pertinent that before launch of the scheme the organization must be ready with all its' schedules and tools like Manual, Circulars, Meeting schedules, Seminar Schedule, Agenda, Printing of Stationery, List of delegations, Opening Ceremony schedule, Medals, Trophy, and allocation of cash etc.

Modus Operandi

Manual

First of all, the suggestion scheme manual suitable to the organization must be formulated / prepared / defined properly in details illustrating categories of suggestions, co-ordinating agencies, evaluating procedure, implementation guideline, reward schemes, documentation, authority and responsibility and reports etc.

Organization / Structure

As per the guideline laid, the co-ordinating agencies including team members are selected through out the organization at all level. They are provided with the copy of working manual for day-to-day reference.

STRUCTURE (SC)

LEVEL-1: (AT ORGANIZATION LEVEL):-

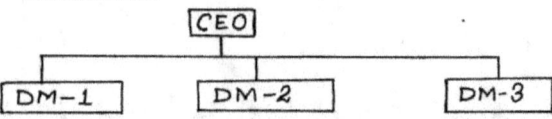

LEVEL-2: (AT DIVISION LEVEL):-

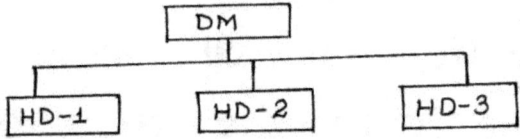

LEVEL-3: (AT DEPARTMENT LEVEL):

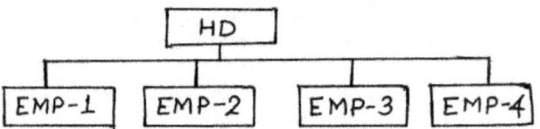

NOTE :-

    EMP - EMPLOYEE.
    HD - HEAD OF THE DEPARTMENT.
    DM - DIVISIONAL MANAGER.
    CEO - CHIEF EXECUTIVE OFFICER.
    SC - SUGGESTION COMMITTEE

Launching & Invitation

The top management, periodically, releases an invitation / thanks letter to all employee to participate in the suggestion scheme and for their all previous contributions.

Participation

As by default all employee are the member of this scheme and overall organizational growth is the prime factor, all employee should participate.

Generation

Employee generates improvement suggestion on prescribed/preprinted format and submits it to their respective coordinator in the department. In case of space shortage, additional sheets are attached.

SUGGESTION FORM

ABC COMPANY PVT. LTD.

Sl.No.:...1234..........   Date :....../......./....

Topic :...

Existing Scenario...

....

Suggetion..

....

Suggestor's Details :-

Name :..

EMP Code:...

Dept. :..   ------------

Signature

----

counter foil   Sl.No...1234.....

Topic...

Office stamp   Co-ordinator

SUGGESTION FORM

ABC COMPANY PVT. LTD.

Sl.No.:....1234.........   Date :....../......../....

Remarks and Comments ....
....
....
....
....
....
....
....
....

Signature
2/3

SUGGESTION FORM

ABC COMPANY PVT. LTD.

Sl.No.:...1234.........   Date :....../........./....

Cost and Benefit Analysis per annum :-...
....
....
....
....

Benefit Factor :-

$$= \frac{\text{Benefit} - \text{Cost}}{\text{Cost}} =$$

Implemented on Date :....../........./.........   **Signature**

Final Recommendation

Approved for Appreciation / Cash Prize / Increment / Promotion .

--------------   ------------------

Secretary   Vice-President

3/3

Acknowledgment

A receipt/counter foil is provided to contributor by coordinator on all received suggestions after recording/updating on the computer system.

Evaluation

The evaluation of suggestion must be done on a fair basis without any bias and prejudice of any kind by evaluating committee/member. The process of evaluation works

in two ways. It does not only evaluate the suggestion but also provides a learning opportunity to evaluator too.

A few points, mentioned below, are of great importance for the evaluating committee/member.

The suggestion must be clear in its' meaning. Most of the time, the description/details provided does not convey the main idea and purpose. In case of any doubt, the committee must approach contributor and or his senior for clarity on the topic.

Study each suggestion well. Don't jump to conclusions.

Be generous with FIRST – TIME suggestors. Make personal additions to the suggestion, if necessary.

Motivate workers to achieve higher goals.

Consider the level of the suggestor when evaluating his suggestion.

Evaluate suggestions promptly. If delay is unavoidable, notify suggestor the reason for it.

Add a few words of encouragement, especially if the idea is not accepted.

Testing and Prototyping

The evaluated and accepted suggestion is required to be tested on a prototype model and once again it is evaluated for the final output / result.

These results are presented in the suggestion committee at the apex level to get the nod for final implementation.

Implementation

The implementation of each and every suggestion, finally accepted, must be done in a time bound manner .

Appreciation

All participants, not awarded must be provided with appreciation letter for participation.

SUGGESTION SCEME

ABC CO. PVT. LTD.

Appreciation Letter

SL.NO.: 1234   DATE ::...../...../......

This is to certified that Mr./Ms.......................... EMP Code.............has been awarded with this appreciation letter against his/her suggestion .... Dated.../..../...... . We look forward to receive his/her valuable suggestions in future also.

We wish him/her a prosperous future.

Secretary

Reward

Certificate, Cash reward, Growth plan are provided after implementation to suggestor according to the category and benefits of the suggestion.

SUGGESTION SCEME

ABC CO. PVT. LTD.

Certificate

SL.NO.: 1234   DATE ::....../....../......

This is to certified that Mr./Ms.......................... EMP Code..............has been awarded with Cash Prize of Rs.........../ a special annual increment / a promotion against his / her contribution through suggestion .... Dated.../..../........

We wish him/her a prosperous future.

------------   ------------------   ------------
Secretary   Vice-president   President

Recognition

Each and every rewarded suggestor must be recognized by publishing their suggestion in the company magazine / news letter / reports and displaying on the notice boards with photographs.

MAGAZINE (SUGGESTION SCHEME)

NOTE :

W — WINNER

Data Analysis and Presentation

Annual analysis and data presentation is done every year in the following manner as presented in the graphs below to asses the performance of the scheme and take corrective action time to time.

a. Participation per year ::

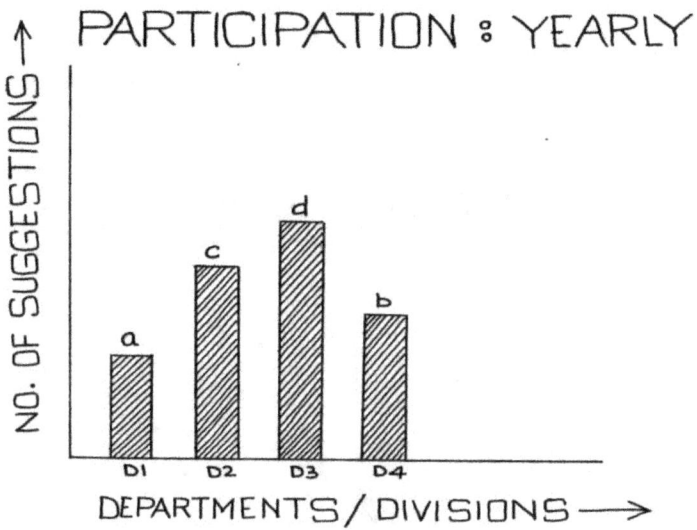

b. Benefits per suggestion per year ::

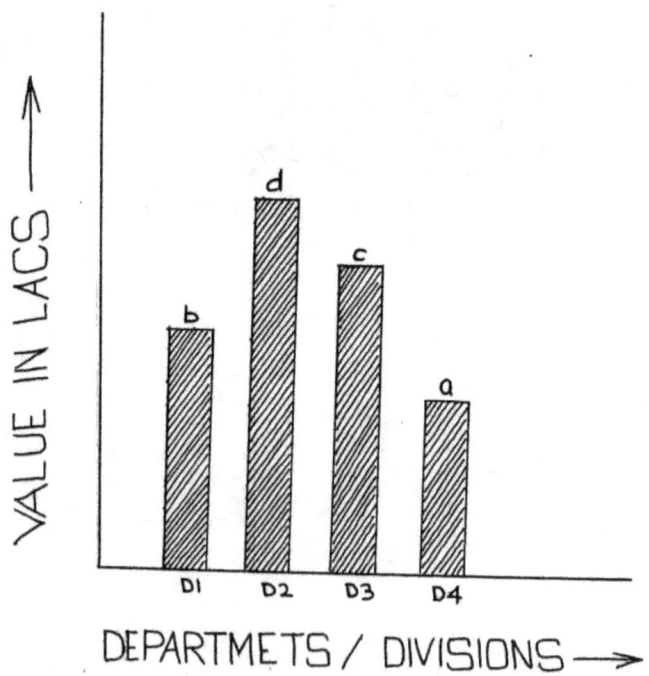

BENEFITS/SUGGESTION/ANNUM

c. Benefits at Organization level per year ::

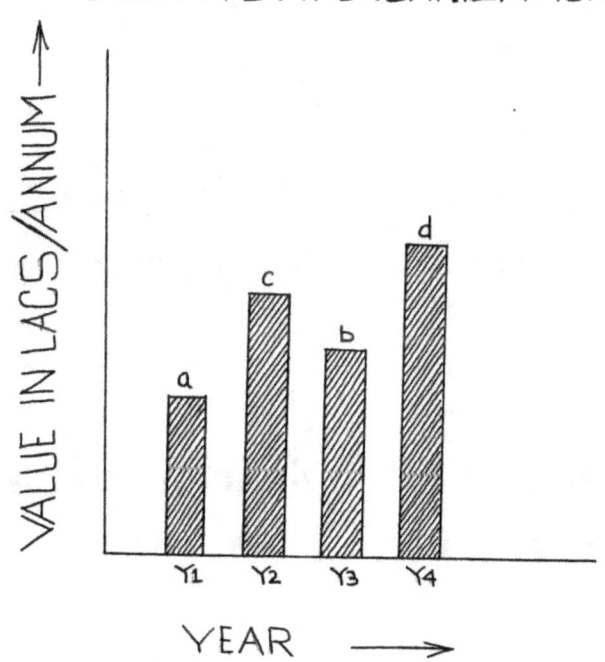

BENEFITS AT ORGANIZATION LEVEL

VALUE IN LACS/ANNUM →

a   c   b   d

Y1   Y2   Y3   Y4

YEAR →

Do's and Do Not's

DO'S

1. Each and every suggestion must be appreciated.

2. Participants must be encouraged irrespective of suggestion weight-age.

3. Set a good example by participating in the scheme.

4. Focus on the frequent problem area.

5. Listen to lower staff and other supervisors and understand what needs to be improved.

6. Provide advice to prepare suggestions in a better way including improving the handwriting.

7. Examine each and every suggestion carefully. In case of any doubt, discuss it with the participant to understand their point of view.

8. All suggestions must be examined from implementing point of view, regardless of the magnitude of the suggestion and participant should be educated for implementation also.

9. We should make the entire process voluntary and no direct or indirect force should be applied. Otherwise, this will only negate their creativity.

10. Encourage them to view the problem from different perspectives and consider what implementation will require. This will increase their understanding of the system improvement.

11. All the participants should be provided with training and seminars covering topics like setting goal , collecting information and implementation steps etc.

12. Train and guide junior staff how to write a suggestion in a better way.

13. Promoting the suggestion system.

14. Establishing communicating policies across the organization.

15. Create an atmosphere where it is easy to make suggestions.

16. Expand the horizon of staff perspective view of thinking.

17. Encourage the positivity among lower staff.

18.   Build confidence in lower staff.

19.   Guide them to observe things closely.

20.   Indicate Problems and educate lower staff to look for problems.

21.   Solve a problem together, analyze and develop solutions.

22.   Make lower staff to understand the process fully.

23.   Focus on problems that affect the lower staff themselves.

24. Give lower staff examples of good suggestions for study.

25. Let employee attend programme on creativity and self improvement.

26. Provide with a chance to meet with employee of other departments who are also involved in making improvements.

27. Focus on the problems such as 1. Required number of suggestion not met. 2. Quality of suggestion does not improve. 3. Lacks enthusiasm 4. Good ideas poorly written. 5. Lower staff not finding satisfaction in suggesting and improving.

28. Ask what inconveniences the lower staff has experienced in working.

29. Make an improvement on employee job and write the improvement on a suggestion form as a sample.

30. Have the "new" employee and a more experienced employee work together to identify a problem and suggest improvement.

DO Not's

1. Do not ridicule any participant. This create fear and discourages participants and hence reduction in number of suggestions.

2. Avoid using words like 1. everyone knows this. 2. it has never been tested. 3. We have tried out this before. 4. It won't work. 5. It is impossible to implement. 6. This is not complete. 7. You have not planned in a proper way. 8. Why do you want to change this when it is working fine. 9. There are rules to be followed. 10. Technically not fit. 11. Boss would not like it. 12. Your idea

is farfetched. 13. We do not have the budget for your good idea. 14. It may create problem later. 15. Do not come to me for advice. 16. Suggestion is not clear make it somewhat better.

****

A Few Examples

A list of few examples of suggestions / topics of probable suggestions are being furnished below. This is not the exhaustive list. This should be taken as reference only.

1. Rationalization of activities / materials / processes / clients / customers / partners / equipments / products / documents / forms etc.

2. Automation of Manual Activities.

3. Removal of redundant Activities and assets.

4. Reduction in movement of men / machines / materials.

5. Security measures for data / information / assets / products / men / machines / equipments / materials / products.

6. Office Automation like attendance / reduce usage of hardcopy / digitization / reduce Manual work etc.

7. Introduction of specific tools and gadgets for quality Improvement as per industry.

8. Suggestion for organizational changes for improvement.

9. Reduction of resources like man / machine / material.

10.   Reduction in process time by eliminating or merging steps.

11.   Requirements of training / retraining / on the job training.

12.   Up-skilling.

13.   Products simplification / making user friendly.

14.   Cost reduction.

15. Efficient man power utilization.

16. Quality Improvement in the areas of data, information, man, machine, material, product , tools, methods.

17. Products design Improvement.

18. Highlighting Safety concern.

19. Security measures of physical assets, raw data/information, documents, design etc. from internal and external sources.

20. Better work organization.

21.   Focusing on Family welfare.

22.   Reducing Health hazards.

23.   Creation and use of Standard Operating Procedure.

24.   Implementation of Group technology.

25.   Introduction of Automated Guided Vehicle.

****

APPENDIX

Activities

Author

Automation

AGV

Cash

Certificate

Client

Cost

Customer

Digitization

Eliminating

Equipment

Examples

Family Welfare

Form

Group Technology

Growth

Health Hazard

Improvement

Job

Machine

Magazine

Man

Materials

Medal

Merging

Method

News Letter

Organization

Partner

Plan

Power

Product Design

Reduction

Resources

SOP

Suggestion

Suggestion Scheme

Tools

Training

Trophy

Utilization

Work by Author

Work Organization

✳✳✳✳✳✳✳✳✳✳✳✳✳✳✳✳✳✳

WORK BY AUTHOR

The below furnished list is not comprehensive.

BOOKS

Seven Tools

5 S

Swarlipi

Suggestion Scheme

Filmy Geet (A Collection)

Ghazal (A Collection)

Compendiums(Mathematics)

Sankalan (A Collection from Literature)

The Author

"The suggestion scheme is a boon to all organizations irrespective of size, and or product. Its' benefits are beyond imagination and run parallel to the scheme throughout the lifespan of organization itself. "

I, D. C. Prasad (Dinesh Chandra Prasad), am a mechanical engineer with wide experience in various fields like Industries, Teaching, Training, Insurance, reading & writing, music and Computer.

My sincere thanks is due to all my family members, relatives, teacher and GURU (in school and college) and all seniors (at work places and training centers) whose effort, support and contributions have played an important role in my development to this level to-day.

I will be grateful to have suggestions from my readers to improve this book.

Dinesh Chandra
dcprasad.dcp@gmail.com